# Nutritional Management After Bariatric Surgery

**Chrisanthy Vlachaki**

*Clinical Dietician-Nutritionist, B.Sc., M.Sc.*

**Athens, 2008**

**ISBN Number : 978-0-615-24857-8**

**Publisher : Chrisanthy Vlachaki**

## Table of Contents

# CHAPTER 1 – Introduction

Obesity represents a multifactorial disease that not only has a genetic inclination but also depends on several environmental factors. The rising prevalence of morbid obesity (body mass index ≥40 kg/m$^2$) is contributing to wide-ranging health problems worldwide and is now estimated to affect one in 50 adults [1].

Patients that suffer from pathogenic obesity have to deal with physical as well as mental problems, not to mention the increasing levels of morbidity and mortality. Of course, the initially chosen therapy is a non-interventional programme that has to do with the alteration of patients' nutritional habits, and includes intense physical activity and psychological support joint with pharmaceutical treatment. Nevertheless, in most cases this kind of therapy has only temporary effects. Moreover weight regain after treatment discontinuation is often significant [2, 3]. The outcomes may improve the health and psychosocial status of people with moderate obesity but probably have little effect on the health and psychosocial status of those with clinically severe obesity.

While behavioural modification and/or pharmacotherapy (principally orlistat and sibutramine) have so far failed to cause significant sustained weight loss, bariatric surgery can result in sustained weight loss, as well as striking remission of many obesity-related co morbidities for most morbidly obese individuals [4, 5].

## CHAPTER 2 – Surgical methods

Since 1950, several surgical methods have practiced in the therapy of pathogenic obesity [6]. Bariatric operations undoubtedly represent the most effective method for a long-term maintenance of important weight loss in pathogenic obese patients (Body Mass Index > 40 or Body Mass Index > 35 with at least two obesity implications) [7, 8]. For the majority of patients, the advantages of weight loss through the metabolic syndrome therapy far outweigh the risks of postoperative complications, which are familiar and can easily be anticipated or dealt with [9]. Loss of extra weight of about 40-80% has mentioned according to the type of operation [10]. Moreover, numerous studies support that the improvement in patients' life quality is very important [11-14]. Thus, bariatric surgery has become a more popular treatment option for those with clinically severe obesity or those with BMI > 35 in the presence of significant obesity related co morbidities.

There are two categories of bariatric operations:

1. Restrictive, and
2. Combination of restrictive and malabsorptive operations.

Nutritional needs and dietetic treatment varies according to the grade of restriction or malabsorption that is caused by the operation, as well as by the specific region of the intestine that is bypassed. The various nutritional elements are generally absorbed from a specific area of the digestive system. Weight loss is accomplished with a reduction in food intake, or a reduction in the absorption of the food from the intestine, or finally with the combination of the above methods.

There are clues which support that weight loss after bariatric surgical methods is made from normal alterations in the secretion of gastric hormones, nevertheless the exact mechanism is not yet totally understandable [15-17]. This is the way that the basic goal is promoted, which basic goal is weight loss, and diminished energy intake and absorption accompanied by diminished intake and absorption of necessary nutrients. Consequently, deficiencies in nutritional elements are possible in case that the postoperational nutritional instructions are not strictly followed, and in case that the nutritional stage of the patient is not evaluated.

It is of extremely importance due to undernutrition imposed on the patient and inherent complications after all bariatric procedures, that the patients should be monitored by a group of specialists that is scientists specialized in obesity issues [18]. The multidisciplinary team with medical, surgical, psychiatric, and nutritional expertise is important in both pre- and post-operative periods [18].

## 2.1 Restrictive operations

Restrictive operations, like a) vertical banded gastroplasty, b) adjustable gastric band and c) sleeve gastrectomy, have the aim to restrict the total volume of food intake and finally accomplish weight loss. This goal is reached by a reduction in the volume of stomach, which as a result leads to a feeling of repletion after the intake of a small amount of food. In a short term patients must be aware of the fact that these operations demand very good chew and deceleration in the rate of food consumption. Failure in the alteration of these consuming habits could lead to frequent vomits or dysphagia [19, 20]. Rarely, the patients that fail to alternate their habits result in experiencing vomits, and run the risk of developing protein-energy malnutrition, due to absolute absence of food [21].

### 2.1.1 Vertical banded gastroplasty

Vertical banded gastroplasty creates a vertical gastric sac of 30 ml volume, in the entrance of stomach and along the upper section of minor arrow. This sac communicates with the rest of the stomach through an orifice of 10-15 mm diameter, which is amplified by a grid. This method causes an average weight loss of about 50% for the first 3 years. However, after a long-term observation of patients, an important average of failure (over 50%) was established in which patients retrieved their weight after about ten years. Moreover, at least 30% of patients have to undergo another operation in order to narrow the gastric sac or due to insufficient weight loss. Vertical banded gastroplasty is not appropriate for obese people that consume a lot of sweets. This operation is not usually applied nowadays.

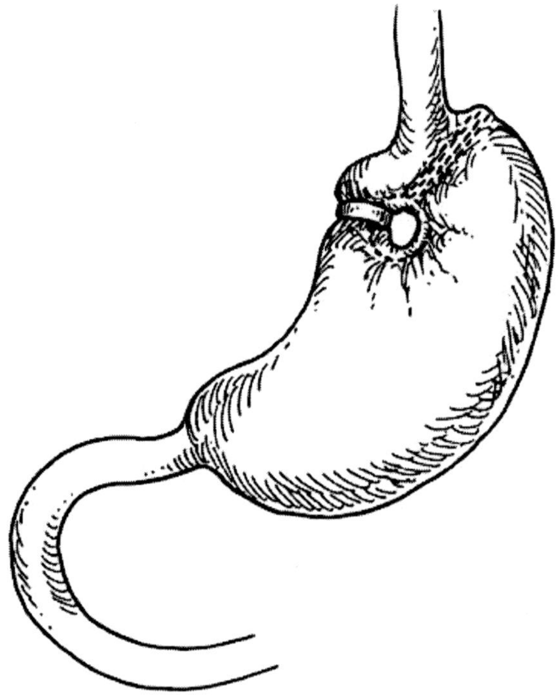

Figure 1. Vertical banded gastroplasty

### 2.1.2 Adjustable gastric band

Gastric band is placed laparoscopically in the entrance of the stomach. The major advantage of this method is that the band has the ability to be adjusted, so that the size of the orifice that food passes through, change according to the patient's tolerance. For example, in case a patient experiences frequent vomits, the band is adjusted so as to permit the easier pass of the food to the stomach. In reverse, if a patient can consume more food than it is aimed, the band is adjusted in a way that his diameter will be reduced. Consumption of food is reduced, and adequate weight loss is accomplished.

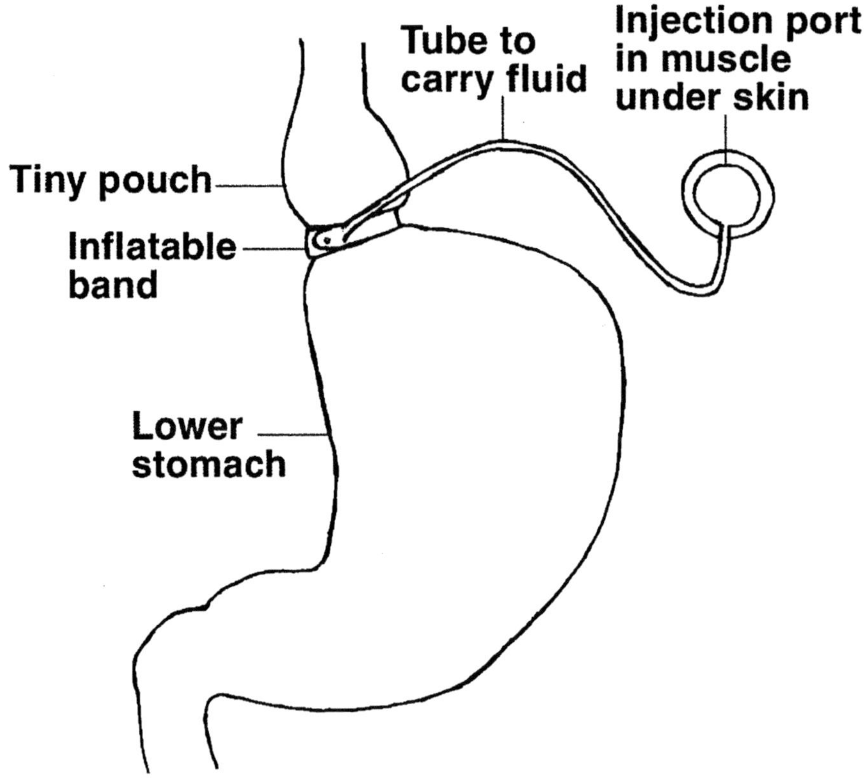


Figure 2. Adjustable gastric band

### 2.1.3 Sleeve gastrectomy

Sleeve gastrectomy is the newest surgical technique in the treatment of pathogenic obesity. During operation, patient's stomach reduces significantly in size. The new stomach has now a tubular shape, like a "sleeve". With this technique the quantity of food that can be taken is limited. Moreover, in the stomach the production of ghrelin, a hormone that causes an increase on the feeling of hunger, is reduced. The stomach is limited by junctures in order to receive 100-150 ml of food, while the pylorus that regulates the normal exit of food to the duodenum and the small intestine stays intact. After the operation the patient looses the extra weight in two ways. Firstly, the quantity of food is reduced due to the small size of the stomach. Secondly, the patient looses his appetite for consumption of large amount of food. This is a laparoscopic operation that lasts about 90 minutes, and has very good results even in extra-obese people with body mass index higher than 50.

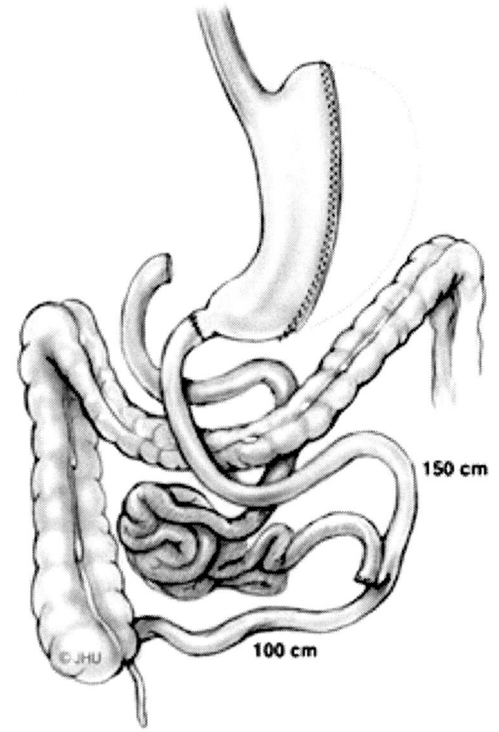


Figure 3. Sleeve gastrectomy

## 2.2 Combination of restrictive and malabsorptive operations

The operations of restrictive and malabsorptive type comprise of: a) Roux-en-Y gastric bypass, b) biliopancreatic diversion, in which a part of the stomach is removed (restrictive operation) and at the same time a big part of the small intestine is bypassed (malabsorptive operation), and c) biliopancreatic diversion with a duodenal switch, which is like biliopancreatic bypass, that is the remove of stomach with duodenal switch that regulates the rate of evacuation of the remaining stomach. These operations are popular in the United States of America, as well as in other places of the world [22].

### 2.2.1 Roux-en-Y gastric bypass

Roux-en-Y gastric bypass includes the bypass of the main body and the cavity of the stomach, the duodenum, and a part of jejunum [23]. The new stomach that receives the food, has 30ml volume (that is quantity equal to 2 tablespoons). This fact has as a result a reduced rate of consumption and absorption of the nutritional elements. The amount of malabsorption is defined by the surgeon and the length of the small intestine that is bypassed during the operation. A frequent phenomenon that is observed after this operation is the Dumping syndrome.

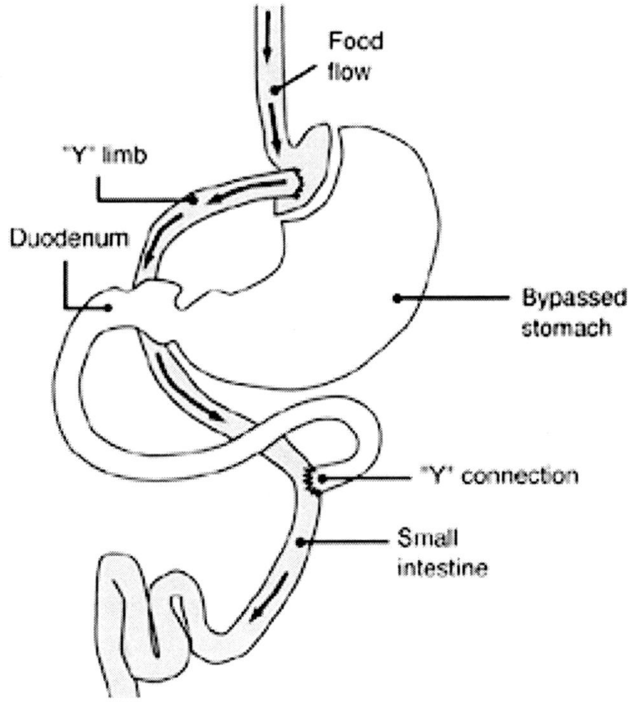


Figure 4. Roux-en-Y gastric bypass

### 2.2.2 Biliopancreatic diversion

Biliopancreatic diversion is executed as a restrictive gastrectomy, which is continued with a bypass of the small intestine, and through a Roux-en-Y inosculation, a common intestinal channel of 50 cm length is created, and this is the part that the food is absorbed. The patient can eat larger quantities of food than in the Roux-en-Y gastric bypass, but because of the significant malabsorption, aqueous, malodorous bowel and gases are caused [24]. Protein malnutrition is a possible complication of this operation, which is observed in about 15% of patients [25]. When this complication is observed the patient may need totally parenteral nutrition for 2-3 weeks. The malabsorption of fat and fat-soluble vitamins is also important. Deficiencies in the fat-soluble vitamins A, D, E and K have been observed [26]. The supplements that are essential for patients with Roux-en-Y gastric bypass, are also essential for patients with biliopancreatic diversion in a higher dosage, moreover the dispensation of supplement of D vitamin is vital in these patients.

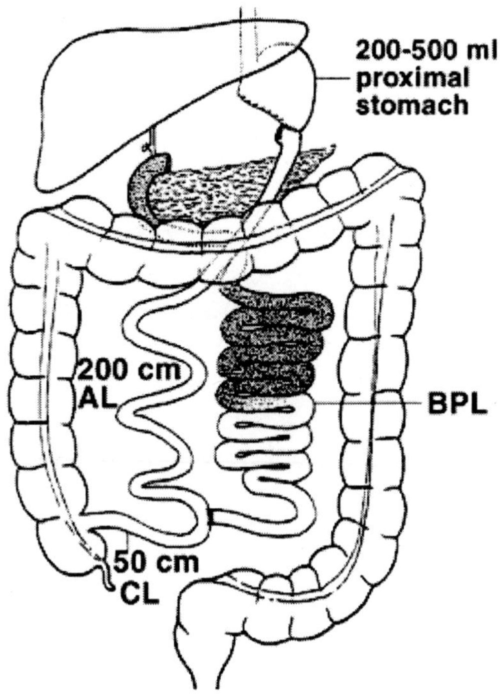


Figure 5. Biliopancreatic diversion

### 2.2.3 Biliopancreatic diversion with a duodenal switch

Biliopancreatic diversion with a duodenal switch is a combined restrictive and malabsorptive operation. The length of the common intestinal channel that absorbs the nutritional elements is 75-100 cm or 125 cm. The protein malnutrition is observed at a percentage 3% to 5% of the population. The malabsorption of fat is highly important, so vitamins A, D, E and K are required in water-soluble form [24]. It is useful that patients should be trained to have a diet low in fats, in order to reduce the production of aqueous evacuation. A supplement of iron as well as calcium is suggested. The absorption of iron and calcium is improved from this surgical method, as compared to the rest malabsorptive operations, due to the 5 cm of duodenum that are retained [24]. When the duodenum is retained the supplement of vitamin B12 is not essential.

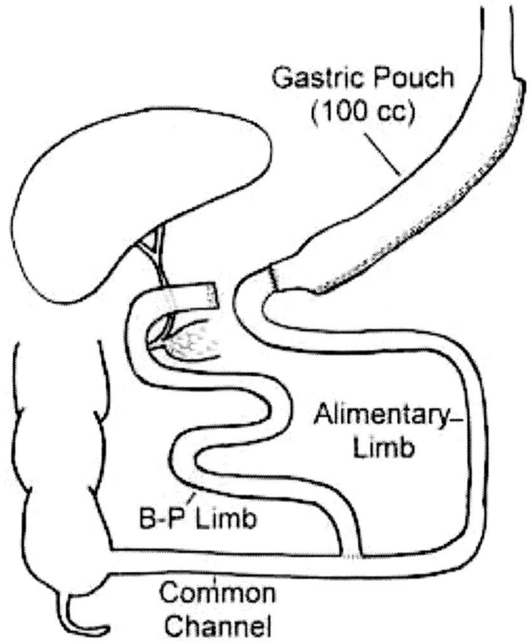


Figure 6. Biliopancreatic diversion with a duodenal switch

## CHAPTER 3 – Preoperational nutritional aspects

The decision to operate a patient results from an analysis by a multidisciplinary team, bringing together the surgeon, the dietician and the psychologist. In this evaluation the team must analyze the patient's motivation and ability to keep to the therapeutic program in the long-term. The whole clinical, somatic and psychological situation of the patient must be evaluated very well. The team must study the social and family context of the patient. The patient must be fully informed about the follow-up conditions, all the advantages and disadvantages of each operation and the consequences of the procedure in his daily life.

To be a candidate for bariatric surgery, patient should meet all of the criteria below, as established by the National Institutes of Health:

- Body Mass Index (BMI) of 40 or more or a BMI of 35 to 39.9 with serious medical conditions related to obesity that would improve with weight loss
- Have attempted (and failed) previous weight loss efforts with diet, exercise, lifestyle changes, or medications
- The patient should understand the possible risks, benefits, and side effects of the procedure

- Must understand and be committed to the lifestyle changes necessary to succeed
- Must not have any medical, psychiatric, or emotional condition that would prohibit surgery
- Must be motivated and have realistic expectations of the surgery. Surgery is a tool to lose weight, it is not the magic bullet
- Patient's age should be between 18-65 years of age

Before the surgery it is important that the patient try to lose weight (2-3kg), especially in the abdominal area where the surgery takes place. The patient should be educated to chew very well, to eat slowly and to avoid high-energy foods before surgery. The patient can try to avoid drinking liquids during meals. The dietician can educate patient to recognize foods that contain high amounts of protein, such as meat, poultry, fish, eggs and dairy products, so that he can emphasize on consuming these foods after the bariatric surgery.

# CHAPTER 4 – Postoperational nutritional treatment

Patients, who have undergone malabsorptive operations, run the risk of developing macronutrient deficiencies due to reduced protein intake or absorption, as well as fat malabsorption. Micronutrient deficiencies after bariatric surgery can also occur, such as deficiency in iron, vitamin B12, folic acid, calcium, thiamine, zinc, vitamin A and other fat soluble vitamins [27-29].

## 4.1 Protein-calorie malnutrition

Severe malnutrition is the most serious metabolic complication of bariatric surgery. Patients usually poorly tolerate red meat. If they do not consume enough protein sources, such as milk, yogurt, eggs, fish and poultry, they are good candidates for protein malnutrition. Decreased intake and absorption of protein postoperatively can occasionally lead to protein-calorie malnutrition, especially after malabsorptive procedures [30]. A minimum of 60 to 70 grams of protein per day should be ingested [31, 32]. Protein supplements are often useful in meeting this guideline, but caution is needed because these supplements contain a lot of calories. Many high-protein supplements also contain a lot of calories, since most protein malnourished patients have been or are currently experiencing unwanted or excessive weight loss. For this reason, protein supplements made especially for bariatric patients are preferred. These supplements are high in protein and low in sugar and calories [33]. In cases of severe malnutrition, enteral or parenteral nutrition therapy may be necessary. Dieticians should monitor frequently their patients, especially those who are prone to developing protein-calorie malnutrition.

## 4.2 Fat malabsorption

The more malabsorptive procedures can significantly impair fat absorption, which can result in deficiencies of fat-soluble vitamins [34]. It is evidenced by steatorrhea, which is the principal means by which biliopancreatic diversion, promotes weight loss. Small amount of fat (approximately 10 gr) at each meal should be encouraged to assist with maintaining gallbladder emptying and minimizing the risk of cholelithiasis, especially during rapid weight loss. This is also recommended in patients who have had a cholecystectomy, to limit the symptoms that are associated with fat malabsorption.

### 4.3 Iron

Iron is a mineral that is involved in the oxidation and reduction activities of cellular aspirations. It plays a vital role on red blood cell function, myoglobin activity, and the utilization of multiple heme and nonheme enzymes. Bariatric surgeries that alter the gut physiology and bypass portions of the small intestine compromise digestion and absorption of iron leading to possible iron deficiency. It is a frequent phenomenon that is caused by reduced intake of food rich in iron. Due to the small size of the gastric pouch, many patients have difficulty tolerating meats containing heme iron, the most bioavailable form. Moreover the reduced gastric acid does not permit the alteration of the divalent iron to the absorptive form of trivalent iron. Iron is absorbed in the duodenum that is bypassed. The risk for iron deficiency is increased in proportion to the time passed from the operation, due to the reduction in the iron supplies [34]. International bibliography suggests the systemic dispensation of iron supplement. The dosage fluctuates from 40 to 65 mg of iron per day, especially in pro-climacteric women [28]. This generally averts iron deficiency in most people [35]. Bibliography mentions a frequency of iron deficiency of about 33% to 50% of the patients [29, 36]. Some professionals suggest a dispensation of iron supplement only in case the patient develops iron deficiency anaemia. Nevertheless, there is a possibility even for patients that take iron supplements daily, to develop anaemia [35]. Multivitamin supplementation is not preventive, necessitating supplemental liquid or chewable iron preparations [36]. Iron and ferritin should be measured and deficiencies corrected pre- and post-operatively for the prevention of iron deficiency. Iron sources such as ferrous sulphate, gluconate, or fumarate are usually recommended daily along with added vitamin C. Vitamin C enhances iron absorption due to its role in the conversion from the ferric to the absorbable ferrous form. Iron supplements containing vitamin C are ideal for patients to prevent having to take additional supplements.

### 4.4 Folic acid

Folic acid deficiency is less frequent than iron, vitamin B12 and calcium deficiency, but it can develop after reduced intake of food and therefore food rich in folic acid, like green foliar vegetables, beans, liver, tuna, salmon, bananas, and avocado. Moreover, even if the duodenum, that is the initial region of absorption of the folic acid, is bypassed, the absorption of the folic acid is accomplished along the almost whole small intestine, which has underlain an adjustment after the operation. It is suggested that patients should intake 800 to 1000 mg of folic acid daily [19, 35]. Maintaining adequate folic acid levels is important, because of the possibility of megaloblastic anemia.

### 4.5 Vitamin B12

The absorption of vitamin B12 is significantly reduced after combined operations [36, 37]. The intake of foods that are a good source of vitamin B12, are consumed in little quantities, as the red meat and dairy products are not well tolerable [38]. In a less acidic environment compared to a normal stomach, vitamin B12 is difficult to be released from the foods. Vitamin B12 must be joined with the intrinsic factor-IF in order to form the complex IF/B12 that is absorbed from the ileum [39, 40]. The intrinsic factor is produced in the area of the stomach that is bypassed, so the complex IF/B12 is not accomplished and the absorption is the least possible. Vitamin B12 deficiency is observed on about 33% of the patients that have undergone a Roux-en-Y gastric bypass [41]. Multivitamin supplementation alone does not prevent vitamin B12 deficiency, which can occur in up to one-third of patients [42, 43]. Oral supplementation with 350-600 μg/day is necessary to normalize levels in 95% of patients [41].

### 4.6 Calcium

Morbidly obese individuals may have abnormal serum calcium and vitamin D levels even prior to surgery. This can possibly be explained due to reduced physical activity, less sunlight exposure and increased storage in body fat with reduced bioavailability [44]. Calcium deficiency is frequent, and the metabolic disease of bones represents a possible long-term risk that is associated with the combined operations. Usually, calcium levels in blood are among normal limits, unless other causes induce abnormal rates. If the calcium levels are low, bones will release calcium in blood flow in order to normalise the calcium levels in the serum. After a malabsorptive operation, the patient's bones continually release calcium, in order to maintain normal levels of calcium in the serum. After a time period, osteopenia is observed, and usually the patient detects the problem only when a bone or a tooth breaks. On this point the problem is major. Calcium deficiency is frequent and can generally be averted with the consumption of the right form and the right quantity of calcium. Calcium citrate is the necessary form, since it does not need acid to break it in order to be absorbed [45]. Most supplements contain calcium carbonate, which is not efficient after the Roux-en-Y gastric bypass [46, 47]. In a meta-analysis of calcium bioavailability, Sakhaee et al. concluded that calcium citrate is better absorbed by approximately 22% to 27% [45]. Current recommendations for daily calcium range from 1000 to 1500 mg/day [6, 35, 47].

### 4.7 Thiamin (Vitamin B1)

Thiamin absorption requires acidification of food, and that is why gastric bypass significantly impairs absorption. But thiamin deficiency has also been reported following purely restrictive gastric procedures [48, 49, 50]. The development of thiamin deficiency, known as Wernicke encephalopathy or cerebral beriberi, is a rare but potentially fatal long-term complication of bariatric surgeries. Symptoms of Wernicke encephalopathy include ataxia, drowsiness and confusion. It is caused mainly due to drastically decreased intake and excessive vomiting. Onset is usually 2-4 months after surgery when liver thiamin stores are depleted, but symptoms can present as soon as 6 weeks postoperatively [51, 52]. A multivitamin supplement is usually adequate in preventing thiamin deficiency. If deficiency occurs, treatment consists of immediate administration of parenteral thiamin 50 to 200 mg/day, especially before administration of fluids containing glucose [53, 54]. When symptoms clear, treatment consists of 10 to 100 mg by mouth daily.

### 4.8 Vitamin A and other fat soluble vitamins

Deficiencies of fat-soluble vitamins are more common after the malabsorptive procedures. All patients with biliopancreatic diversion should receive supplemental vitamin A and vitamin D and have their serum levels monitored and maintained in the normal range. The recommendation for vitamin D intake is 8 μg/day. Clinical manifestations of vitamin K deficiency have not been published despite a report of high incidence of hypovitaminemia K after biliopancreatic diversion [55]. Vitamin E deficiency is not common in patients supplemented with multivitamin, although it is reported after biliopancreatic diversion [56]. Roux-en-Y gastric bypass patients usually a multivitamin supplement for prevention of deficiency.

## 4.9 Zinc

Zinc is a nutrient that depends on fat absorption. That is the reason, why low serum concentrations were observed in patients after biliopancreatic diversion or duodenal switch [55]. It has been reported zinc deficiency after a Roux-en-Y gastric bypass, which developed persistent postprandial emesis [57]. Although zinc deficiency has not been widely reported, it is not commonly monitored. It may play a role in the generalized alopecia that commonly follows bariatric surgery, although inadequate protein intake and stress may play a role too [51]. There is one report on resolution of alopecia after supplementation of high doses of zinc sulphate in patients submitted to vertical gastroplasty [58].

## CHAPTER 5 – Nutritional stages

Nutritional instructions vary between professionals who have to do with health and depend on the specific operation. Nutrition is initiated solely with liquids. The best choices are water, tea, jelly without sugar, and broth. Initially the above must be consumed in small quantities [59]. If the pylorus has undergone alterations or the duodenum has been bypassed, there is a high risk of Dumping syndrome to develop. The simple sugars that are in a concentrated form must be neglected from the nutrition. Juice is hyperosmotic and has higher concentration in sugar than the non-alcoholic drinks, so juice must not be ranked in clean liquids. If milk is well tolerable, it can be added to the nutrition after 1 or 2 days, in order to provide protein in liquid form. Nevertheless, there are two cases that must be considered according to the milk tolerance, and these are: content in fat, and content in lactose.

If a patient has undergone biliopancreatic diversion with or without a duodenal switch, it would be better for him to be given milk without fat. In Roux-en-Y gastric bypass, milk with 2% fat is appropriate, given that malabsorption of fat is not severe. Lactose, the sugar that milk physically contains, can cause flatulence and gases to people with lactose intolerance. Malabsorptive operations can cause lactase deficiency [60], either suddenly or in a long term, and this fact can lead to diarrhoea, flatulence and gases after the consumption of milk. In these occasions, soy milk or lactose free milk is suggested. These types of milk contain the same quantity of protein as compared with the normal milk, and they do not contain lactose.

Generally, mashed soft food enters the nutrition list for a time period of 2 weeks, about 1 to 2 weeks after the operation. During this period, the size of the meal gradually increases, till the volume that the new created stomach can afford. Some patients claim that they experience an increase in their appetite or the feeling of hunger after 10 to 14 days after the operation. Also, most patients during this period believe that water consuming is very improved and they can consume it easier. It is suggested that the patient should start consuming solid food for another 2 to 4 weeks, after a month that he has left the hospital [61].

The exact promotion rate of the constitution of the nutrition as well as the size of the meal depends on patient's tolerance, the type of the operation, and the experience of the scientific group that monitors the patient. It is important for everyone to know that different health professionals promote the stages of nutrition slowly or faster. Moreover, patients vary according to their readiness to continue to the next stage of nutrition [59]. The recommendations that mentioned above are the most frequent nowadays. Red meat, dry poultry or pork, bread or generally starch, celery, and coconut can cause a number of problems to many patients for a long time period, even for years [24, 59, 62]. Nevertheless, other patients can tolerate every food without facing problems. The most substantial problems are caused by dry food, and food with gummy constitution [59]. Patients get trained so as to add in their diet only one food at a time. This fact enables the determination of foods that are not tolerable.

For a successful weight loss, changes are essential in nutritional behaviour, as well as the prevention of post-operational complications [26, 63]. Patients should follow predetermined instructions according to the size of the meal, in order to avoid overeat. This means that patients should chew their food very well, and shallow it slowly. Liquids must not be consumed in the same time that solids are consumed. Specifically, when there is a gastric band, liquids must be consumed 10 minutes before, or 30 minutes after the meal [64]. For the rest bariatric methods, liquids must be consumed 30 minutes before or 30 minutes after the meal [59, 60]. The consumption of liquids during the meal can cause vomit or diarrhoea and make the patient feel hungry after the meal faster, fact that drives the patient to pickings between the meals. Foods that contain protein must be consumed before foods that contain only carbohydrates or fat [59, 60]. Also, foods rich in calories must be avoided in order to achieve weight loss and intake of foods with high nutritional value [64]. Patients are trained in order to stop the meal when they feel repletion.

## CHAPTER 6 – Management of complications

Nearly all morbidly obese patients with satisfactory postoperative weight loss, experience substantial improvement in the quality of their lives. On the other hand, complications occur frequently, both small and large, just as there is with any surgical procedure.

Complications that may occur include dumping syndrome, cholelithiasis, diarrhea, constipation, dysphagia, dehydration, vomiting. Additional surgeries may be needed to correct any health problems that occur because of these complications, and, of course, death is a risk as well. Patients with a very high BMI and/or severe medical conditions are at the highest risk, but death can also occur in any patient.

Complications can't always be prevented but the risks can be minimized. Early recognition of these complications and attention to details are of great importance in this group of high-risk patients. The surgeon along with the patient will weigh any potential complications against the risks of not having the procedure at all.

### 6.1 Dumping syndrome

Additionally to the distributive differences between the patients according to their tolerance in the constitutions of the foods, sugar and sweets can cause Dumping syndrome. This syndrome is a normal answer to simple sugars. Dumping syndrome is characterized by an array of symptoms which are: a feeling of dizziness and sweating that is followed by a quick cardiac rate and most of times from severe diarrhoea. When sugar is consumed, it passes quickly through the small intestine, causing an osmotic charge, which leads to a liquid transposition from blood to the intestine [33, 35, 64]. The reduction in the blood volume increases the cardiac rate, and as a result the patient feels tired and he wants to lie down (something that improves the cardiac flow). The answer of insulin causes hypoglycaemia symptoms. The inflow of liquids in the intestine, because of the osmotic change, can lead to aqueous diarrhoea. There are two types of Dumping. Early dumping occurs 30 to 60 minutes after eating and can last up to 60 minutes. It occurs as a result of rapid emptying to sugars or carbohydrates from the gastric pouch into the small intestine. Late dumping occurs 1 to 3 hours after eating. It is related to increased insulin after oral glucose with subsequent hypoglycaemia (low blood sugar).

Dumping syndrome is usually a desirable complication, given the fact that it creates a negative experience in sugar consumption. Patients that suffer from Dumping syndrome rapidly learn to avoid sugar. Sensitivity in sugar varies from patient to patient. Some patients may never face the syndrome, even after the consumption of bon-bon, ice cream, or non-alcoholic drinks. Other patients are very sensitive and can suffer from this syndrome even after the consumption of 2gr of sugar. Generally, simple sugars in fresh fruits, dairies and vegetables are well tolerable. Patients must be trained so as to avoid juices, drinks that contain sugar like sports and energy drinks, soda, sweets and must be careful with spices and sauces that contain sugar like ketchup. Gradually, the small intestine fits to the new conditions and tolerates highly osmotic food, so the syndrome is not observed for years after the operation, even if the patient consumes sweets and foods containing fat [61].

## 6.2 Cholelithiasis

Rapid weight loss is frequently connected to cholelithiasis. It has mentioned that 50% of the patients that have undergone gastric bypass were observed with dense bile (like mud) in the gall bladder [63]. Some surgeons decide to remove gall bladder during bariatric operations. The laparoscopic removal of the bile is usually safe and efficient in patients with incidental cholelithiasis without complications. The surgical treatment of cholelithiasis can be complicated due to the difficult access after the bariatric operation [64].

## 6.3 Diarrhea

Diarrhea or loose stools is mainly a potential side effect of biliopancreatic diversion with duodenal switch. It may be seen with Roux-en-Y gastric bypass, but would not be associated with the adjustable gastric banding. Diet is a major influence on bowel movements. Reducing the amount of fat will usually have a direct beneficial effect on the number and quality of bowel movements a patient may have, thus less possibility of experiencing diarrhea. If the patient has lactose intolerance, this can result in diarrhea and gas complaints. An early step in the assessment of diarrhea after surgery is to eliminate dairy products from the diet completely. Many patients will benefit from a course of probiotics. Rarely, with the above nutritional management and pharmaceutical treatment, a patient may continue to have an unacceptable high frequency of bowel movements. It is reasonable to give the bowel time to compensate and adapt, so that fat absorption increases. This will allow diarrhea to improve.

### 6.4 Constipation

Constipation is also seen in patients after bariatric surgery. It is usually due to insufficient intake of water, and may be corrected by attention to water ingestion, addition of fiber products and avoidance of diuretics such as caffeine, which decrease the amount of water in the body.

### 6.5 Dysphagia

Dysphagia is a medical term for difficulty swallowing. This may be a side effect of any operation that causes restriction. This is caused by eating too fast, too much or not chewing well enough when the stomach has been made smaller.

This is the reason why the food backs up into the oesophagus and causes chest pressure or even tightness in the throat. It is important to educate patients in order to give time when they are eating. They must stop eating and drinking, if they have dysphagia. Dysphagia may result in vomits. It can be avoided if the patient chews well (approximately 20-30 times every bite). It can also help avoiding tough foods, such as doughy bread, overcooked steak or dry chicken.

## 6.6 Dehydration

Dehydration is commonly seen after gastric restrictive intestinal bypass surgeries and is attributed mainly to decreased intake [59]. Frequent vomiting or diarrhea may exacerbate dehydration. Obese patients require greater amounts of fluid than their nonobese counterparts to maintain normal fluid balance and homeostasis.

It is recommended that patients drink 1 litre of fluids per day for the first two weeks after surgery. Since the patient can't drink large amounts of liquids at one time, it is recommended that he sips fluids frequently throughout the day and that he does not drink large gulps. For some patients the recommendation of 1 litre per day may be a very large amount, because it is common that some patients do not feel thirsty after surgery. So patients should not be advised to drink fluids only if they are thirsty. They have to be educated to remember to drink fluids all day.

This amount should be increased to 2 litres per day. All liquids should be consumed before or after meals. Following a schedule is important in order to prevent dehydration, dumping syndrome and vomiting.

Caffeinated beverages such as coffee and tea should be avoided, because caffeine increases urine output, which can increase risk for dehydration. When hydration is adequate, caffeine may be consumed in moderation.

### 6.7 Vomiting

Vomiting almost always occurs during the first few months after surgery. It typically happens one to three times a week and it is caused by overeating, not chewing food adequately and swallowing big food pieces [59]. Patients need time to adjust to the smaller gastric pouch that is made after surgery (restrictive or combination bariatric surgery). Vomiting is well tolerated by most patients but if it becomes frequent, nutritional deficiencies may occur, such as low potassium or low magnesium levels. If vomiting is frequent, oral supplementation of these elements is required [33]. Vomiting can be associated with non-diet related problems such as strictures or stomal stenosis. If this happens, the patient should be referred as quickly as possible to rule out surgical complications [59].

## CHAPTER 7 – Other important instructions

Diet in every advanced phase must contain foods rich in protein, as the major components of every meal. Protein is the priority for every operation, especially in biliopancreatic diversion with or without duodenum switch. Large intestine, as time passes, fits and begins to absorb protein, so protein deficiency is generally improving through time.

Liquid consumption can be difficult for the first 2 weeks, basically due to oedema caused by the operation. Patients must be trained to consume at least 1 litre of liquids every day, and increase this quantity to 2 litres when they feel able for a quantity like this.

Generally, a supplement of vitamins and minerals, preferably in liquid form or dispensable, is enough after a restrictive operation, given the fact that the phenomenon of malabsorption of specific nutritional elements does not exist. There is a significant reduction in the quantity of the consumed food, and therefore a significant reduction in the quantity of the sum of micro-nutritional elements. For this reason a supplement that will restore the levels of these elements in the RDA (Recommended Dietary Allowances) levels, is required. A chewable vitamin is suggested 2 times a day for the first month after the operation in order to cover the most of needs in vitamins and minerals. In case Wernicke-Korsakoff syndrome and peripheral neuropathy will be observed, an additional dispensation of the vitamins A, E, B1, B6 and B12 for the first 12-16 weeks is required. This phenomenon is generally rare and after vomits.

Patients must decrease the consumption of fats and avoid foods that are low in nutritional value. When the solid diet is established, and liquid consumption is sufficient, drinks that contain lots of calories must be forbidden, given the fact that these drinks will decelerate weight loss. Carbonated drinks must be avoided as carbon dioxide can cause discomfort. Also, caffeine must be avoided till the consumption of 2 litres of liquids, because of it's diuretic properties.

## CHAPTER 8 – Total parenteral nutrition

The complication of the operation, and the poor health condition of the patients before the operation, can drive the patient to a critical condition that requires special care. It is totally wrong for someone to consider that a pathogenic obese person eats well [62]. These people have a large stock of energy in the form of fat, but some of them have reduced levels of protein in their body, due to their nutrition which is poor in nutritional value. Every post-operational complication can cause the catabolism of the non-fatty body mass. As in every person who has normal weight, an obese patient must begin with a total parenteral nutrition, in case he has not eaten for more than 7 days [65].

It is fact that an obese patient could have deficient nutrition till the day before the initial operation. If the access through the intestine is feasible after the bariatric operation, this is the way that must be firstly chosen for the nutrition of the patient [65]. Each one of the professionals must be in high alert in case refeeding syndrome is observed after the bariatric operation [28]. The levels of potassium, phosphorus, and magnesium can be significantly reduced, given the fact that they follow the intracellular movement of glucose, and this can cause a probably dangerous fall in these micro-nutritional elements. It is generally suggested that the initial infusion of total parenteral nutrition, for the first 24 hours, will begin with 50% of the considered needs of the patient and 50% of the considered liquid volume. After 24 hours, the total perenteral nutrition can be increased in goal with the full nutritional needs. Potassium, phosphorus, and magnesium must be checked daily for 3 days. Low levels of these elements must be supplemented and normalized immediately.

Generally, after a bariatric operation, patients' needs in calories, in order to retain their weight, must not be satisfied by calories that come from fat and dextrose. International bibliography frequently suggests undercaloric nutrition with efficient intake of protein [65]. It is also suggested that non-protein calories should be calculated to about 15 to 20 kcal/kg of the corrected weight of the body. The corrected weight of the body is calculated from the following type: current weight-ideal weight*25%+ideal weight=corrected weight. Many times, due to high charge of dextrose that is dispensed, and the frequency of diabetes or the non-tolerance in glucose on the obese population, insulin must be added to the total parenteral nutrition or be given in a proportional scale, in order to regulate the increased levels of glucose in the blood.

Protein is calculated to 1-2gr per kilogram of corrected body weight. The corrected body weight takes into account the added non-fatty body mass that a person has, in order to lift the extra weight. When calories and protein are calculated, based on the current weight, needs are overestimated, because fatty body mass is metabolically inactive, and the amount of non-fatty body mass, as percentage of the total body weight, is low when compared with a person who have a normal body weight. Fats must never be calculated in a dosage over 1gr of fat per kilogram of corrected body weight [65]. Micro-nutritional elements (vitamins and minerals) are calculated in a standard dosage unless the condition of the patient requires another treatment. The patient must be informed that weight loss may decelerate when total parenteral nutrition is used, but it is essential treatment for the health restoration after the operation.

## CHAPTER 9 – Weight loss and exercise after bariatric surgery

In general, the malabsorptive procedures lead to more weight loss than the restrictive procedures. A meta-analysis [5] reports the following weight loss at 36 months:

- Biliopancreatic diversion - 53 kg
- Roux-en-Y gastric bypass (RYGB) - 41 kg
    - Open - 42 kg
    - Laparoscopic - 38 kg
- Adjustable gastric banding - 35 kg
- Vertical banded gastroplasty - 32 kg

An exercise program will tone muscles and assist with weight reduction and total body fitness. The patient should obtain physician approval before starting an exercise program. A daily exercise program is encouraged for most patients. The dietician can give some tips on how to get started. The exercise physiologist will set up a more individualized program.

## CHAPTER 10 – Reduced mortality and morbidity

Several recent studies report decrease in mortality and severity of medical conditions after bariatric surgery [66, 67, 68]. In the Swedish prospective matched controlled trial, patients with a BMI of 34 or more for men and 38 or more for women underwent various types of bariatric surgery and were followed for an average of 11 years. Surgery patients had 5.0% mortality while control patients had 6.3% mortality. In a Utah retrospective cohort study that followed patients for an average of 7 years after various types of gastric bypass, surgery patients had 0.4% mortality while control patients had 0.6% mortality [67].

A randomized, controlled trial in Australia compared laparoscopic adjustable gastric banding with non-surgical therapy in 80 moderately obese adults (BMI 30-35). At 2 years, the surgically-treated group lost more weight (21.6% of initial weight vs. 5.5%) and had statistically significant improvement in blood pressure, measures of diabetic control, and HDL cholesterol [68].

## CHAPTER 11 – Bariatric surgery for adolescents

Rates of obesity among youth are on the rise. Bariatric surgery is sometimes considered as a treatment option for adolescents who have developed extreme obesity. Although it is becoming clear that adolescents can lose weight following bariatric surgery, there are numerous unanswered questions about the long-term effects of these operations on adolescents.

Adolescents being considered for bariatric surgery should:

- Have failed ≥6 months of available, organized attempts at weight management, as determined with the assistance of their primary care provider
- Have attained or nearly attained physiologic maturity
- Be very severely obese (BMI ≥40 kg/m$^2$) with serious obesity-related comorbidities or have a BMI of ≥50 kg/m$^2$ with less severe comorbidities
- Demonstrate commitment to comprehensive medical and psychological evaluations both before and after surgery
- Agree to avoid pregnancy for at least 1 year postoperatively
- Be capable of and willing to adhere to nutritional guidelines postoperatively
- Provide informed assent to surgical treatment

- Demonstrate decisional capacity
- Have a supportive family environment

Over the years, gastric bypass surgery has been the predominant operation used to treat adolescent extreme obesity. An estimated 2,700 adolescent bariatric surgeries were performed between 1996 and 2003 [69]. As yet, the adjustable gastric band has not been approved for use in the United States for people younger than age 18, but favorable weight-loss outcomes following adjustable gastric banding for adolescents have been reported internationally.

# CHAPTER 12 – Conclusion

Nutritional and metabolic complications can be frequent after bariatric operations. The type of the operation usually indicates which deficiencies are expected. There is a possibility for macro- and micro-nutritional deficiencies, which in some cases can be so severe that set the patient's life in danger. Generally, the training of the patient is essential, as well as the continual monitoring and evaluation of patient's nutritional condition. The dispensation of supplements for deficiencies' prevention is crucial.

Despite the evidence of nutritional deficiencies, many health professionals do not recommend adequate supplements or they do not evaluate serum nutrient levels. On the other hand, most patients, who are not well educated on their nutritional attitudes after bariatric surgery, do not comply with the recommended supplement regimen.

That is the reason why education of both health professionals and patients is necessary to prevent malnutrition after bariatric surgery. Bariatric surgery has a place in treatment of carefully selected, well educated patients with a good follow-up.

## References


1. Sturm R. Increases in clinically severe obesity in the United States, 1986-2000. *Arch Intern Med* 2003, 163:2146-8.
2. Davidson MH, Hauptman J, DiGirolamo M, et al. Weight control and risk factor reduction in obese subjects treated for 2 years with orlistat: a randomized controlled trial. *JAMA* 1999, 281:235-42.
3. Wadden TA, Sternberg JA, Letizia KA, Stunkard AJ, Foster GD. Treatment of obesity by very-low-calorie diet, behaviour therapy and their combination: a five-year perspective. *Int J Obes Relat Metab Disord* 1989, 51:167-72.
4. Snow V, Barry P, Fitterman N, et al. Pharmacologic and surgical management of obesity in primary care: a critical practice quideline from the American College of Physicians. *Ann Intern Med* 2005, 142:525-31.
5. Maggard MA, Shugarman LR, Suttorp M, et al. Meta-analysis: surgical treatment of obesity. *Ann Intern Med* 2005, 142:547-59.
6. Deitel M, Shikora S. The development of the surgical treatment of morbid obesity. *J Am Coll Nutr* 2002, 21:365-71.

7. Van Hee RHGG. Biliopancreatic diversion in the surgical treatment of morbid obesity. *World J Surg* 2004, 28:435-44.

8. Jones KBJ. Bariatric surgery – where do we go from here? *Int Surg* 2004, 89: 51-7.

9. Malinowski SS. Nutritional and Metabolic Complications of Bariatric Surgery. *Am J Med Sci* 2006, 331:219-25.

10. Fisher BL, Schauer P. Medical and surgical options in the treatment of severe obesity. *Am J Surg* 2002, 184:9S-16S.

11. Marinari GM, Murelli F, Camerini G, et al. A 15-year evaluation of biliopancreatic diversion according to the bariatric analysis reporting outcome system (BAROS). *Obes Surg* 2004, 14:325-8.

12. Dittmar M, Heintz A, Hardt J, et al. Metabolic and psychosocial effects of minimal invasive gastric banding for morbid obesity. *Metabolism* 2003, 52:1551-7.

13. Ballantyne GH. Measuring outcomes following bariatric surgery: weight loss parameters, improvement in co-morbid conditions, change in quality of life and patient satisfaction. *Obes Surg* 2003, 13:954-64.

14. Wolf AM, Falcone AR, Kortner B, Kuhlmann HW. BAROS: an effective system to evaluate the results of patients after bariatric surgery. *Obes Surg* 2000, 10:445-50.

15. Faraj M, Havel PJ, Phelis S, Blanke D. Plasma acylation stimulating protein, adiponectin, leptin, and ghrelin before and after weight loss induced by gastric bypass surgery in morbidly obese subjects. *J Clin Endocrinol Metab* 2003, 88:1594-602.

16. Korner J, Bessler M, Cirilo LJ, et al. Effects of Roux-en-Y gastric bypass surgery on fasting and postprandial concentrations of plasma ghrelin, peptide YY, and insulin. *J Clin Endocrinol Metab* 2005, 90:359-65.

17. Cummings DE, Weigle DS, Frayo S, et al. Plasma ghrelin levels after diet-induced weight loss or gastric bypass surgery. *N Engl J Med* 2002, 346:1623-30.

18. Marcus JD, Elkins GR. Development of a model for a structured support group for patients following bariatric surgery. *Obes Surg* 2004, 14:103-6.

19. Miskowiak J, Honore K, Larsen L, Anderson B. Food intake before and after gastroplasty for morbid obesity. *Scand J Gastroenterol* 1985, 20:925-8.

20. Andersen T, Larssen U. Dietary outcome in patients treated with a gastroplasty program. *Am J Clin Nutr* 1989, 50:1328-40.

21. Mason EE. Vertical banded gastroplasty for obesity. *Arch Surg* 1982, 117:701-6.

22. Palmer JA, Marliss EB. The present status of surgical procedures for obesity. In: *Nutrition in Clinical Surgery*. Baltimore, Md: Williams & Wilkins, 1980, 281-92.

23. Sugerman HJ, Kellum JM, Engle KM, et al. Gastric bypass for treating severe obesity. *Am J Clin Nutr* 1992, 55:560s-6s.

24. Marceau P, Hould FS, Simard S, et al. Biliopancreatic diversion with duodenal switch. *World J Surg* 1998, 9:947-54.

25. Scopinaro N, Gianetta E, Friedman D, et al. Evolution of biliopancreatic bypass. *Clin Nutr* 1985, 5:137-46.

26. Mason ME, Jalagani H, Vinik AI. Metabolic complications of bariatric surgery: diagnosis and management issues. *Gastroenterol Clin North Am* 2005, 34:25-33.

27. MacClean LD, Rhode BM, Shizgal HM. Nutrition following gastric operations for morbid obesity. *Ann Surg* 1983, 198:347-55.

28. Kushner R. Managing the obese patient after bariatric surgery: a case report of severe malnutrition and review of the literature. *J Parenter Enteral Nutri* 2000, 24:126-32.

29. Halverson JD. Micronutrient deficiencies after gastric bypass for morbid obesity. *Am Surg* 1985, 52:594-8.

30. Marinari GM, Murelli LB, Camerini G, et al. A 15-year evaluation of biliopanceatic diversion according to the

Bariatric Analysis Reporting Outcome System (BAROS). *Obes Surg* 2004, 14:325-8.

31. Stocker DJ. Management of the bariatric surgery patient. *Endocrinol Metab Clin North Am* 2003, 32:437-57.

32. Marcason W. What are the dietary guidelines following bariatric surgery? *J Am Diet Assoc* 2004, 104:487-8.

33. Fujioka K. Follow-up of nutritional and metabolic problems after bariatric surgery. *Diabetes Care* 2005, 28:481-4.

34. Avinoah E, Ovnat A, Charuzi I. Nutritional status seven years after Roux-en-Y gastric bypass surgery. *Surgery* 1992, 111:137-42.

35. Elliot K. Nutritional considerations after bariatric surgery. *Crit Care Nurs Q* 2003, 26:133-8.

36. Brolin, R. E. Gastric bypass. *Surg Clin North Am* 2001, 81, 1077-95.

37. Amaral J, Thompson W, Caldwell M, et al. Prospective hematologic evaluation of gastric exclusion surgery for morbid obesity. *Ann Surg* 1984, 201:186-93.

38. Institute of Medicine and National Academy of Sciences. *Dietary reference intakes for thiamin, riboflavin, niacin, vitamin b6, folate, vitamin b12, pantothenic acid, biotin, and choline*, 1998.

39. Smith CD, Herkes SB, Behrns KE, et al. Gastric acid secretion and vitamin B12 absorption after vertical Rouxen-Y gastric bypass for morbid obesity. *Ann Surg* 1993, 218:91-6.

40. Behrns KE, Smith CD, Sarr MG. Prospective evaluation of gastric acid secretion and cobalamin absorption following gastric bypass for clinically severe obesity. *Dig Dis Sci* 1994, 39:315-20.

41. Rhode BM, Tamin H, Gilfix B, et al. Treatment of vitamin B12 deficiency after gastric surgery for severe obesity. *Obes Surg* 2005, 5:154-8.

42. Brolin RE, Gorman JH, Gomran RC, et al. Are vitamin B12 and folate deficiency clinically important after Roux-en-Y gastric bypass? *J Gastrointest Surg* 1998, 2:436-42.

43. Kushner R. Managing the obese patient after bariatric suergery: a case report of severe malnutrition and review of the literature. *JPEN J Parenter Enteral Nutr* 2000, 24:126-32.

44. Wortsman J, Matsuoka LY, Chen TC, et al. Decreased bioavailability of vitamin D in obesity. Am J Clin Nutr 2000, 72:690-3.

45. Sakhaee K, Bhuket T, Adams-Huet B, et al. Meta-analysis of calcium bioavailability: a comparison of calcium citrate with calcium carbonate. *Am J Ther* 1998, 6:313-21.

46. Charles P. Calcium absorption and calcium bioavailability. *J Intern Med* 1992, 231:161-8.

47. Goode LR, Brolin RE, Chowdhury HA, et al. Bone and gastric bypass surgery: effects of dietary calcium and vitamin D. *Obesity Res* 2004, 12:40-7.

48. Sola E, Morillas C, Garzon S, et al. Rapid onset of Wernicke's encephalopathy following gastric restrictive surgery. *Obes Surg* 2003. 13:661-2.

49. Bozbora A, Coskun H, Ozarmagan S, et al. A rare complication of adjustable gastric banding: Wernicke's encephalopathy. *Obes Surg* 2000, 10:274-5.

50. Chaves LC, Faintuch J, Kahwage S, et al. A cluster of polyneuropathy and Wernicke-Korsakoff syndrome in a bariatric unit. *Obes Surg* 2002, 12:328-34.

51. Kulkarni S, Lee AG, Holstein SA, et al. You are what you eat. Surv Ophthalmol 2005, 50:389-93.

52. Escalona A, Perez G, Leon F, et al. Wernicke's encephalopathy following bariatric surgery: clinical course and MRI correlation. Obes Surg 2004, 14:129-32.

53. Primavera A, Brusa G, Novello P, et al. Wernicke-Korsakoff encephalopathy following bioliopancreatic diversion 1993, 3:175-7.

54. Beers M, Berkow R, eds. The Merck Manual of Diagnosis and Therapy. Whitestation, NJ: Merck Research Laboratories, 1999.

55. Slater GH, Ren Cj, Siegel N, et al. Serum fat-soluble vitamin deficiency and abnormal calcium metabolism after malabsorptive bariatric surgery. *J Gastrointest Surg* 2004, 8:48-55.

56. Vazquez C, Morejon E, Munoz C, et al. Nutritional effect of bariatric surgery with Scopinaro operation. Analysis of 40 cases. *Nutr Hosp* 2003, 18:189-193.

57. Brolin RE, LaMarca LB, Kenler HA, et al. Malabsorptive gastric bypass in patients with superobesity. *J Gastrointest Surg* 2002, 6:195-203, discussion 4-5.

58. Neve HJ, Bhatti WA, Soulsby C, et al. Reversal of hair loss following vertical gastroplasty when treated with zinc sulphate. *Obes Surg* 1996, 6:63-5.

59. Parkes E. Nutritional Management of Patients after Bariatric Surgery. *Am J Med Sci* 2006, 331:207-13.

60. Shils ME. *Modern Nutrition in Health and* Philadelphia, Pa: Williams & Wilkins, 1999.

61. Favretti F, O'Brien PE, Dixon JB. Patient management after LAP-BAND placement. *Am J Surg* 2002, 184:38S-41S.

62. Brolin R. Bariatric surgery and long-term control of morbid obesity. *JAMA* 2002, 288:2793-6.

63. Shiffman ML, Sugerman HJ, Kellum JM, et al. Gallstone formation in patients with morbid obesity: relationship to body weight, weight loss and gallbladder bile cholesterol solubility. *Int J Obes Relat Metab Disord* 1993, 17:153-8.

64. Byrne TK. Complications of surgery for obesity. *Surg Clin North Am* 2001, 81:1181-93.

65. Kushner RF, Wall-Alonso E, Alverdy J. Obesity. *A.S.P.E.N. Nutrition Support Practice Manual*. Silver Spring, Md: Merritt RJ American Society Parenteral and Enteral Nutrition, 1988, chap. 21.

66. Sjöström L, Narbro K, Sjöström CD, et al. Effects of bariatric surgery on mortality in Swedish obese subjects. *N Engl J Med* 2007, 357(8):741–52.

67. Adams TD, Gress RE, Smith SC, et al. Long-term mortality after gastric bypass surgery. *N Eng. J Med* 2007, 357(8):753–61.

68. O'Brien P, Dixon J, Laurie C, et al. Treatment of Mild to Moderate Obesity with Laparoscopic Adjustable Gastric Banding or an Intensive Medical Program. *Annals of Internal Medicine* 2006, 144:625–43.

69. Tsai W, Inge T, Burd R. Bariatric Surgery in Adolescents: Recent National Trends in Use and In-Hospital Outcome. *Arch Pediatr Adolesc Med* 2007, 161:217-221.

CPSIA information can be obtained at www.ICGtesting.com
Printed in the USA
BVOW051133210612
293334BV00001B/5/P